Julian Traublinger

Boat Refugees in the Mediterranean

Tackle the Root Causes or Build Fortress Europe?

Anchor Compact

Traublinger, Julian: Boat Refugees in the Mediterranean: Tackle the Root Causes or Build Fortress Europe? Hamburg, Anchor Academic Publishing 2013
Original title of the thesis:

Buch-ISBN: 978-3-95489-098-9
PDF-eBook-ISBN: 978-3-95489-598-4
Druck/Herstellung: Anchor Academic Publishing, Hamburg, 2013
Additionally: Universität Salzburg, Salzburg, Österreich, Bachelorarbeit, Januar 2008

Bibliografische Information der Deutschen Nationalbibliothek:
Die Deutsche Nationalbibliothek verzeichnet diese Publikation in der Deutschen Nationalbibliografie; detaillierte bibliografische Daten sind im Internet über http://dnb.d-nb.de abrufbar

Bibliographical Information of the German National Library:
The German National Library lists this publication in the German National Bibliography. Detailed bibliographic data can be found at: http://dnb.d-nb.de

Es ist nicht die Einwanderung, die Probleme schafft – es ist der politische Umgang damit. Es ist auch nicht Europa, das gestürmt wird – es sind die Mauern einer Festung, die derzeit gebaut wird.

It is not the immigration, which causes problems – it is how it is dealt with it politically. It is also not Europe, which is becoming attacked – however, it is the walls of a fortress, which is being built at the moment.

Corinna Milborn in „Gestürmte Festung Europa...“, p.7, 2006

Abstract

In this study about the boat people in the Mediterranean an emphasis lies on Italy and further on the EU policy on refugees in the Mediterranean under cooperation with African states. Regarding the EU policy it is especially about the border security agency Frontex that has available a rapidly growing budget coming from the EU funds.

The protagonists of the EU politics like to stress that their policies are for the best of the migrants themselves as well as their countries of origins and the ones they are going to, so that they could produce a "triple win situation". But refugee aid and other human rights organizations criticize harshly that refugees are criminalized and do not get their legitimate rights guaranteed. The NGOs collected numerous examples of inhumane policies against "illegal" migrants in countries around the Mediterranean.

By bringing together the current events, the author wants to describe state networks about illegal migration and follow the question if there is the wish of the EU to accord the refugees their rights at all.

Table of Contents

Tables

Illustrations and Pictures

Abbreviations

ASL	Associazione Sanitaria Locale, Health Service
CDI	Centro d'identificazione, Identification Center
CIR	Consiglio Italiano per i Rifugiati, Italian Refugees Council
CPA	Centro di Prima Accoglienza, Center for the First "Welcome"
CPT	Centro di Permanenza Temporanea e Assistenza, Refugees' Prison
EC	European Community
ECRE	European Council on Refugees and Exiles
ed.	Editor
EMN	Europäisches Migrationsnetzwerk, European Migration Network
EPN	European Patrols Network
EU	European Union
f.	and following page
ff.	and following page and page after the following
Frontex	European Agency for the Management of Operational Cooperation at the External Borders of the Member States of the European Union
GUS	Gemeinschaft Unabhängiger Staaten, Community of Independent Countries
ICF	Consorzio Italiano della Solidarietà, Italian NGO, counseling Refugees
IMF	International Monetary Fund
MP	Member of Parliament
MSF	Medici senza Frontieri, Doctors without Borders
PNA	programma nazionale asilo, National Programme for Asylum seekers
PRODEIN	Asociación Pro Derechos de la Infancia, Child Rights Organisation
Sciences Po	Institute for Political Studies
UIS/UIL	Italian Workers' Union
UNHCR	United Nations High Commissioner for Refugees
UNTOC	United Nations Convention against Trans-national Organized Crime
WTO	World Trade Organization

Research Question

How does the EU and especially Italy cope with the illegal Immigration of refugees over the Mediterranean Sea? How does the political position of the EU influence the refugees' problematic situation and which (other, better) possibilities of policies would exist to help refugees, further legal immigration and hinder illegal immigration with the loss of many lives?

Hypothesis

The position of the EU is to rather defend against refugees than anything else. Furthermore it strengthens its forces at its borders so to build the often described „Fortress Europe". A substantial solution for the refugee problem is not in sight. There is no purpose to help the "illegal" migrants, more the opposite. On the search for solutions the EU concentrates on the strengthening of the border control first of all in the first place and less on long term strategies to fight the problem at its roots e.g. through development cooperation.

Introduction

The EU countries Spain, Italy, Greece and Malta as Mediterranean countries do have heavy problems through the immigration of refugees on boats at their coasts. Especially Spain cannot cope with the arrivals of refugees that come over the Atlantic Ocean to its overseas territory Canary Islands. The fate of the migrants, who die at high numbers in the sea, is a constant tragedy of an unimaginable scale.

In this study about boat people at the Mediterranean Sea I am going to emphasize on Italy and furthermore the EU policies about refugees in the Mediterranean area under cooperation with African states. Concerning the EU policies I am going to tell something about the EU border surveillance agency Frontex which has got a fast growing budget from EU funds[1].

The actors of the EU polity like to stress that their policies are best for the migrants themselves and their countries where they come from and where they are heading to. Therefore they call it a "triple-win-situation"[2]. But refugee aid and other human rights organisations criticize harshly that refugees would become criminalised and would not have the possibility to enjoy their rights.[3] The NGOs collected numerous examples of a policy not respecting human beings regarding "illegal" immigrants in Mediterranean countries. Through this study I want to describe governmental networks, thus through collecting the present time incidents. Furthermore I want to clarify the question if it is a goal of the European countries at all to let the refugees have their rights.

Boat People – Policies of Italy and the EU

In the main part of this study I am going to explain in chapter 1 the push and pull factors that have influence on migrants from African countries heading to the Mediterranean area. Those are reasons for exhausting foot trails through the North African Dessert and the crossing of the sea of crowds in boats not built for the sea. This is the topic of chapter 2. In chapter 3 I am going to stress a core topic of this study namely the policy of Italy concerning the boat people. The last big chapter (No. 4) goes about the EU policy with the border patrol agency Frontex and the case of Libya as subtopics. The terms boat people, (illegal) migrants and (illegal) refugees are used nearly synonymous here. The reason for me is that I did not see myself able to conduct a clarification of terms within the framework of this study. As well as I doubt about the sense of such a categorization. This in practice would mostly be difficult anyway.

---

[1] see Fortress Europe 2007, p.3, from the internet, filed under: http://www.statewatch.org/news/2007/nov/fortress-europe-libya-report.pdf
[2] see Rede des deutschen Innenministers Wolfgang Schäuble am 19. November 2007, from the internet, filed under: http://www.bmi.bund.de/nn_662956/Internet/Content/Nachrichten/Reden/2007/11/BM__EuroMed.html
[3] see Gleitze / Schultz 2006, pp.16-22, from the internet, filed under: http://www.proasyl.de/fileadmin/proasyl/fm_redakteure/Broschueren_pdf/Zonen_der_Rechtlosigkeit_04.pdf

1. Push and Pull Factors

Push factors means reasons of migration that lie in their country of origin. In opposition pull factors are reasons that make a country attractive for immigrants.

1.1 Problems in the Countries of Origin

Bade made a general statement and was telling firstly about population growth reasons in combination with economic ones. He sees a "Gap between high population growth rates and stagnating or even shrinking job offering rates". Bade particularly sees the agriculture sector being in a rising crisis partly for reasons of globalization. From "growing slums of the `Mega Cities` to where the rural population flees new movements [escape] again".

As further reasons for migration in the developing countries beneath others Bade names desertification through "climatic and [man made] destruction of the environment", political conflicts "that partly occur out of indigenous developments and partly colonial respectively postcolonial conditions" as well as civil wars that "[in] the nineties [were becoming] the most important reasons for refugee movements in the `Third World` [...]".[4]

Bade furthermore draws the attention to the fact that "the panicking Europe is only concerned with about 5 percent" of the "worldwide movements of estimated 120 million people". There would be "namely a gapping difference between global dramas within the refugees world and European stage play with the setting of `being concerned` through a rising worldwide `pressure by migration`, which is estimated in quite different ways." The history of the south north migration towards Europe would be "at the bottom line, rather a history of being afraid of it and of defence against it than a history of migration movements."[5]. This gives somehow evidence to my hypothesis, which says that on behalf of the government they see the refugees as enemies who they need to fight. However, I wonder about Bade's opinion nowadays looking at the rising numbers of the boat people that head towards Europe since the year 2000. The documentary film "Au clair de la lune [...]" from Leona Goldstein about western African countries of origin of refugees in Europe underlines the direction of Bade's statement whereas the entire number of refugees differs. There would currently be 40 million people fleeing worldwide. About 32 to 34 million of them stay within their home countries. Less than 200,000 people were applying for asylum in Europe in 2006 of whom meanwhile 5 percent were able to get recognition.[6]

Texeire from the Science Po in Rennes, France counts as main countries of origin in Africa in

---

[4] see Bade 2000, pp.10ff.
[5] the same, p.13
[6] see in the appendix « 'Au clair de la lune...' - … » … transcription of subtitles

the cases of heading to Malta: Eritrea, Djibouti, Ethiopia, Somalia, Sudan, Congo and Liberia. About Africa she states in general that corruption and a lack of democracy in the former colonised countries mean severe problems. Many politicians had used their powerful positions to fuel ethnical conflicts that had been made worse or even been triggered in the colonial time[7]. Furthermore Texeire says that false governmental policies and corruption had been the reason for starvation catastrophes and a system of distribution that is not feasible to give enough nutrition for the survival of the people. HIV/AIDS she named in addition. As countries with severe human rights violations she counts: The Democratic Republic of Congo, Sierra Leone, Liberia, Sudan and the Ivory Coast[8]. An example for human rights violation on a higher political level can be the murder of Norbert Zongo. In the cited documentary of Leona Goldstein Zongo is shown as "main speaker of the intellectual opposition movement [comment: in Burkina Faso] and editor of the weekly newspaper `L'independent`". It is told that he was stopped on his way home and burnt in his car at the 13th of December 1998 investigating in a murder case in which the brother of the president of Burkina Faso was said to be involved.[9]

The Austrian journalist Milborn explains the reasons of fleeing the country Burkina Faso in depth and states that "only a minimal part" of the eight million "Burkinabé" are heading to Europe[10]. As further countries of origin "at the borders of Europe" Milborn names: Mali, Senegal, Togo, Niger, Guinea Bissau[11]. She writes that European countries "as former colonial and profit winning power and trigger of poverty in Africa [...] [are] quite involved in the causes that make Africans emigrate."[12] As well as she writes about the attractiveness of the immigration countries: "The enormous differences between the rich Europe and the misery in Africa are offered daily on TV in the café around the corner."[13]

Within the pull factors Milborn writes explicitly about the climate change, problems in connection to the growing of cotton especially combined with the phenomenon of the debt trap for such countries. With the example of the village Tolo in the north of Burkina Faso she shows how the countryside in its district dries down. The rainfall becomes rarer and in single cases heavier whereby it gets drier overall and the torrential rain falls destroy the fields in addition. A high tide caused a hole in the dam near the village[14]. Burkina Faso emphasizes the

---

[7] see Texeire 2006, pp.21ff., from the internet, filed under: http://www.alternattiva.org.mt/filebank/documents/Fiona%20TEXEIRE-%20At%20the%20Gate%20of%20Fortress%20Europe.pdf
[8] the same, p.24
[9] see in the appendix « 'Au clair de la lune...' - ... » ... transcription of subtitles
[10] Milborn 2006, p.209
[11] the same, p.195
[12] the same, p.195
[13] the same, p.210
[14] the same, pp.195ff.

cotton growing sector too much following the orders of the IMF and the Worldbank so that the last trees are cut and pesticides are used. This lasts as long as the country is rotten and only dessert remains. Instead of growing corn as their own food base the farmers favour to grow the export ware cotton. In addition the US penetrates the world market with its subsidised cotton through WTO regulations so that the price for cotton in Burkina Faso goes below the production cost.[15] During 2005 there was a big starvation catastrophe in the area where Tolo lies[16]. – "Many young men left the village to earn the money for their families elsewhere."[17]

Milborn took Burkina Faso as an example for a country that is not damaged through a war in the recent history, but nevertheless makes clear how Europe (the EU is besides the US the strongest power within Worldbank, IMF and WTO) is involved in the reasons of emigration from the countries of origin.

1.2 Europe's Attraction

It is easy to grasp that the welfare continent Europe has got an immense attractive power in the eyes of African people who suffer from the problems I described above. Surely some illusions that are caused by the TV or emigrated relatives are vanishing into the air after the arrival in the immigration country. And who of the emigrated ones wants to show the impression that the decision about to emigrate had been wrong and that the new home country is not a good place to stay. However, the migrants have got at least the possibility to earn money and help their relatives with cash transfers as long as they are not hindered from it, because they end up as prostitutes or are bound to other forms of exploitation. "Everyone at the border knows that he or she finds a job in Europe within a few weeks: Some entire economic sectors depend on the – de facto tolerated – illegal hired illegal immigrants", says Milborn in her book. Furthermore she writes about a "lower class of ten million illegal immigrants", that for example would hold the economic sectors agriculture, construction, restaurants and hotels alive[18]. It may seem exaggerated that those branches are completely dependent on the illegal migrants. Nevertheless it seems to be possible that they give their not too less important contribution and that without them we would maybe have to expect quite some trouble.

---

[15] the same, pp.200f.
[16] the same, p.197
[17] the same, p.196
[18] the same, pp.9f.

The number of illegal workers is of course difficult to estimate. Milborn says that the EU estimates that 500,000 illegal migrants come to the EU every year to work there. In Italy 2 million "clandestine" are living regarding estimations of the government, though here it stays unclear regarding Milborn's book if it is the entire number of persons with an illegal status or the number of illegal workers.[19]

Currle writes that human trafficking and smuggling was becoming big business for the organized crime in Italy[20]. [...] "The illegal hiring of foreigners looks like more or less tolerated", says Currle[21]. The German representative of the UNHCR Stefan Berglund is cited in "Migration und Bevölkerung" 2003 (migration and population 2003) as follows: „The Mafiosi in the world switch to human trafficking." This business would be now more profitable than drug dealing.[22]

Besides the possibility to go to Europe to work there lawfully there are two more possibilities. Either immigration after some relatives already are residents in the EU or the marriage with a EU citizen. All three possibilities are handled stricter and stricter by time says Milborn. About the example of Austria she reports that only high ranking business men have the possibility to come to Austria to work there subsequent to the laws.[23]

So how do refugees movements of Africans towards the Mediterranean look like? In the following chapters I am going to write more about two locations in the focus at the European borders.

---

[19] the same, p.73
[20] Currle 2004, p.281
[21] the same
[22] „Italien: Hilfe bei Grenzsicherung gefordert", in: „Migration und Bevölkerung" 9/2003, from the internet, filed under: http://www.migration-info.de/migration_und_bevoelkerung/artikel/030906.htm
[23] the same, pp.73f.

2. Attack at the „Fortress Europe“

The German refugee aid organisation “Pro Asyl“ counts the following locations as the most important border crossings for refugees in the Mediterranean on its map of Europe:

- Canary Islands and the exclaves Ceuta and Melilla (Spanish)
- Italy (for sure because of the situation in Lampedusa and Sicily)
- The island state of Malta south of Sicily
- Greece (because of the Aegean Islands)[24]

For Italy here a table with the numbers of arrivals of immigrants at the southern Italian coasts:

Region	*Apulia*	*Calabria*	*Sicily*
1999	46.481	1.545	1.973
2000	18.990	5.045	2.782
2001	8.546	6.093	5.504
2002	3.372	2.122	18.225

Table: Arrival of immigrants at the southern Italian coasts

Source: Ministero degli Interni, “Lo Stato della Sicurezza in Italia“, 15/08/2004, here from Puggioni 2006, 179, translated by the author

Through the growing strength of power of armed forces in Apulia the human traffickers had been forwarded to Calabria and later Sicily, mainly to Lampedusa and Pantelleria as Puggioni writes.[25]

2.1 Drama at the Fence: Ceuta and Melilla 2005

In autumn 2005 the attack of thousands of migrants against the controlled borders of the Spanish enclaves Ceuta and Melilla at the Moroccan Mediterranean coast caused much attention from the media in Europe. It seemed as if the refugee problem in the Mediterranean would move nearer into the focus of the European journalism for the first time including the aspect of the boat people. And if since then the topic had been repeated again and again.

Whereas the phenomenon that migrants try to cross over the then three to six meters high fences around the two Spanish cities at the African continent each of a length of 10km was not new. During the first eight months of the year 2005 they counted about 11,000 cases in which someone tried to cross the border at Melilla regarding to the newsletter “Migration und Bevölkerung” which received the data from the Guardia Civil. It was special then that big crowds of up to 500 people together attacked the fences for the first time. The Spanish border police and the Moroccan military were not prepared for that attack of these masses of people. According to the newsletter more than 3,000 African people tried to reach the two Spanish

[24] see Pro Asyl without year: map “Europa Mit den wichtigsten Grenzübergängen“
[25] see Puggioni 2006, p.178

enclaves illegally between the 27th of September and the sixth of October 2005. More than 900 of them had been successful. That period is said to be the time of the most intensive attacks.

Hundreds hurted themselves with the blades at the doubled fence, or because they had fallen down from the fence, or because following persons run them over. 14 people died partly because of wounds from gunfire. The surveillance personnel at the border would use rubber bullets only in exceptional cases according to "Migration und Bevölkerung"[26]. However, Milborn, who was also doing research about those causes, seems to implicitly write that the Moroccan military used lethal ammunition regularly in cases of gunfire on refugees. It shot to death at least a dozen of refugees at the border to Ceuta and Melilla Milborn wrote regarding to newspaper articles.[27] The documentary of Goldstein shows a feature of the Spanish children aid organisation Asociación Pro Derechos de la Infancia (PRODEIN) with a witness' statement about one case concerning grave mistreatments by the Guardia Civil of three Africans of whom one died[28].

After the heavy attack in 2005 there had been also more incidents. According to that the speaker for European affairs of the refugees aid organisation Pro Asyl Karl Kopp reports in his article "Hart an der Grenze" ("firm at the border" or also "near to the limit") that in July 2006 three human beings had been shot to death in Melilla during their attempt to climb over the fence. "There was no enquiry conducted about the deaths at the border between Spain and Morocco" says Kopp in 2007[29]. Spain had established a "new iron curtain" at its border in Melilla reports Kerstin Böffgen after an excursion to Melilla at the beginning of February 2007. Meanwhile a fourth fold defence apparatus has to be crossed from Africa: A fence which is six meters tall, with flexible wings on top so that you have to fall down to the ground as soon as you reach them. Afterwards they fall on the "Sirga" a chaotic element of ropes which was introduced after the incidents in 2005 and causes grave injuries. In Addition they activate tear gas and pepper spray. Afterwards you are confronted with a lower fence, but also built with flexible wings and NATO wire (with Z-blades). And last there is again a six meters tall fence. Deep cuts are not easy to avoid.

As soon as the refugees reach the Spanish territory they are often caught by the Guardia Civil and forced to go back to morocco instantly according to Böffgen. That is a clear violation of the Genf refugees' convention which says that there has to be the right to get an asylum

---

[26] „Migration und Bevölkerung", 9/2005, p.1, from the internet, filed under: http://www.migration-info.de/migration_und_bevoelkerung/archiv/ausgaben/ausgabe0509.pdf
[27] Milborn 2006, p.17
[28] see in the appendix: « 'Au clair de la lune...' - … »… transcription of subtitles
[29] see Pro Asyl (Ed.): "Tag des Flüchtlings 2007", p.32, from the internet, filed under: http://www.proasyl.de/fileadmin/proasyl/fm_redakteure/Archiv/HTdF_Hefte/Webversion_Heft_komplett.pdf

process at the border and also a violation of the Spanish law about foreigners.[30] Milborn writes that a refugee who had reached European territory through climbing over the first fence "normally [is] rejected". She illustrates that with a cause of "Doctors without borders". A patrol of the Guardia Civil caught a Cameroonian between the then two fences on the third of January 2005. The officers had bound his hands to his back, beaten him up until he lost consciousness and thrown him back to the Moroccan side of the fence through a door in the fence.

You can add deportations of refugees to the Saharan desert by the Moroccan authorities to the human rights violations at the border. Milborn reports for example that 1,200 of those that had not made it across the fences during the heavy attack 2005 had been brought to the Algerian border in busses chained to each other in midst of the Saharan desert.

"`Doctors without borders` that followed the convoy found over 200 refugees searching their way through the desert, but many died dehydrated. During several weeks afterwards members of the `Frente Polisario` that fight for the independence of the Sahraouies reported about findings of dead bodies"[31]

In the documentary of Goldstein the migrant Bandou Keita who was repelled at the Spanish border reports that "comrades" had been abandoned in the desert. "There were injured people that got their arms or legs broken and were not able to walk any longer. As soon as they were thrown to the desert they had to wait for their death."[32]

The European Council on Refugees and Exiles ECRE criticizes Spain for its misuse of its old contract with Morocco about welcoming back refugees to repel Sub-Saharan refugees without looking for reasons of their emigration.[33]

---

[30] see the same, pp.40f.
[31] Milborn 2006, p.13
[32] see in the Appendix: « 'Au clair de la lune...' - ... »... transcription of subtitles
[33] see Pro Asyl: Newsletter 105, October 2005, from the internet, filed under: http://www.proasyl.de/de/informationen/newsletter-nr-105/index.html#c821

Picture 1: Fenced and Controlled Frontiers against Migrants of Economic Reasons and Asylum Seekers – "Le Monde diplomatique" / taz Verlags- und Vertriebs GmbH 2006, p. 50

After writing about the sad incidents at the Spanish-Moroccan border (not to name the numerous refugees drowning year for year in the sea near Gibraltar) I am going to write about refugees problems regarding to Lampedusa (Italy) in the next chapter.

2.2 Lampedusa

According to Milborn there is a list of documented deaths of migrants and refugees who died during the attempted crossing of the European borders[34]. For the period from 1993 to April 2005 the total sum of 6,366 deaths is resulting, of which 90 percent had been deaths on sea. It is said that a third of all dead bodies gets found. Those proposed facts would mean that the term "Lethal Trap Mediterranean Sea" (Milborn) is correct.[35] The newsletter "Migration und Bevölkerung" reported in 2003 that each year several hundred people had drowned during the

[34] Unite, from the internet, filed under: www.united.non-profit.nl/pdfs/deathlist2005.pdf (via sea and over land), here according to Milborn 2006, p.43

[35] Milborn 2006, pp.42f.

attempt to move from northern Africa to Spain or Italy. Spain is not the favourite destination any more. Because of the Schengen Aquis the border between Spain and Morocco had become a high security border. Still numerous refugees make it to the Canary Islands, but the main destination would be Italy. The Italian coast guard (Guardia Costiera) would find overcrowded refugees boats daily and often had to rescue them from emergency on sea.

The Italian island Lampedusa south of Sicily and south west of Malta is Europe's southernmost territory and situated nearer to Tunisia than to Italy. Because of that it is one of the most favourite destinations of the boat people in the Mediterranean. Milborn gives the population of the Island with 6,000 inhabitants. Yearly there would be 150,000 tourists at the Island on holidays. 1992 the first 71 boat people were rescued there and brought to firm ground.

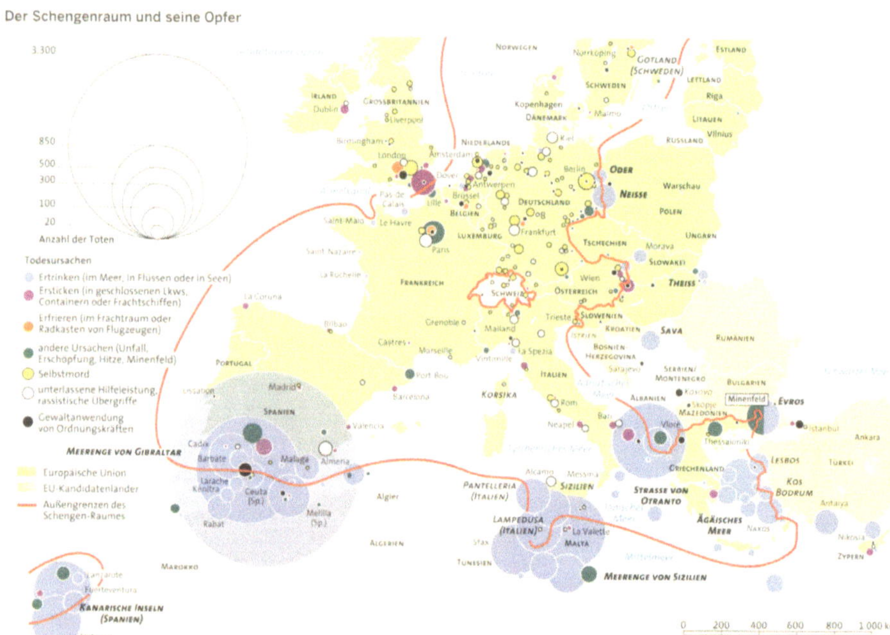

Picture 2: The Schengen Territory and its Victims (blue circles mark deaths by drowning) – "Le Monde diplomatique" / taz Verlags- und Vertriebs GmbH 2006, p.60

Meanwhile a refugee's camp has been established at Lampedusa. It is organized by the Carabinieri and the relief organisation "Misericordia". Neither NGOs nor lawyers, MPs or the UNHCR would have an allowance for entry to the camp as Milborn reports about the situation of those days. The camp is built for 200 people according to Milborn[36], according to

[36] the same, pp.55f.

the newsletter “Migration und Bevölkerung“ of 2003 for 250 people. It would now be the home for 500 people as the newsletter reported in those days. “Because of that the newcomers are brought to the continental territory of Italy immediately.”[37] About the time after the article in the newsletter Milborn reports that hardly any refugee had been leaving the camp in the direction to Italy since October 2004. In that month the first deportation of masses of refugees to Libya had been taking place. Over 1,200 persons had been transported to Libya by plane without proving their reasons of their immigration as asylum seekers. Italy played a trick hereby: Because the boats often are getting fetched in international zones of the sea it would not be a deportation and refusing of asylum, but simply a deportation at the border. Milborn criticizes that hereby it is avoided to tell that someone is allowed to seek for asylum at the border, too, and that numerous of the repelled refugees crossed the border de facto. Because the deportations had been against the Genf convention about refugees the UNHCR was protesting. Nevertheless since then such mass deportations took place regularly and strictly clandestine.[38] In midst of May 2005 there had been taking place rebellions inside the camp because within 24 hours over 1,000 migrants were coming as newcomers and because of that there were created disastrous circumstances especially concerning the hygiene in the camp according to a report in “Migration und Bevölkerung“.[39]

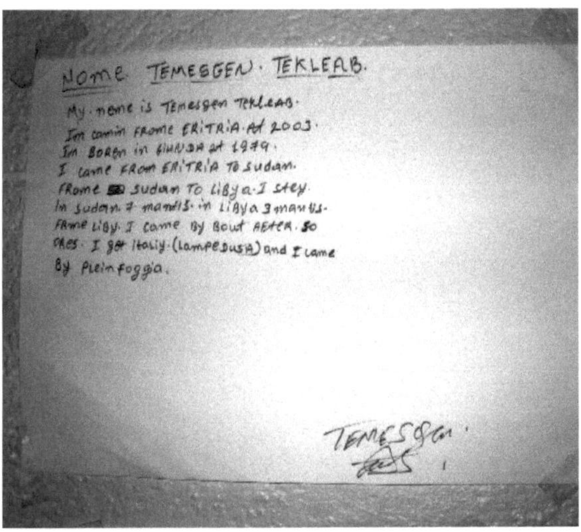

Picture 3: Testimony of a Refugee in Sovereto, Italy[40]

[37] „Mittelmeerraum: Wieder zahlreiche Bootsflüchtlinge gelandet“, from: „Migration und Bevölkerung“ 6/2003, exclusively in the internet, filed under: http://www.migration-info.de/migration_und_bevoelkerung/artikel/030608.htm
[38] Milborn 2006, pp.56f.
[39] „Aufstände auf Lampedusa“, from: „Migration und Bevölkerung“ 6/2005, from the internet, filed under: http://www.migration-info.de/migration_und_bevoelkerung/artikel/050608.htm
[40] Gleitze / Schultz 2006, p.26, from the internet as mentioned above

On the one hand Milborn wrote that MPs had not had permission for entry into the camp at Lampedusa, but at the 28th of June 2005 there must have been a visit of MPs of the EU Parliament to the camp. Judith Gleitze from Pro Asyl cited an email of the EU MP Gabi Zimmer to her. The parliamentarian describes the existing status quo at that time in the camp at Lampedusa as scandalous. Shortly before the visit at least 800 people had been deported from the camp that was built for 190 persons without informing them about the destination. Zimmer reports furthermore that the camp was overcrowded for a longer period before the visit and that some people had to sleep outside on concrete often only with a towel. For to wash oneself there had been only a thin running of water. Although there was a heavy heat the people under detention in the camp had got only a one litre bottle of water two people together. A doctor beneath the MPs had described the supplement of the medical department as disastrous, basic medicine and other medical material had not been existing.[41]

After this description of the status quo at the borders between Europe and Africa with explosive contents I am now going to write about the policy that the illegal migration in Italy is handled with.

[41] „E-Mail der Europaabgeordneten Gabi Zimmer nach dem Besuch im Lager Lampedusa am 28.6.05", from: Flüchtlingsrat Brandenburg: Italia-Info 16.9.2005, without given pages, from the internet, filed under: http://www.proasyl.de/fileadmin/proasyl/fm_redakteure/Newsletter_Anhaenge/105/Italien_info.pdf

3. The Policy of Italy

Italy was a net emigration country after the Second World War until the mid-seventies. Afterwards more people immigrated than emigrated – Italy became an immigration country and during the nineties the main immigration country of southern Europe[42] and had to cope with the situation.

3.1 The Legislature

Currle wrote at the beginning of her comments about main laws in Italy about the "[structural] contradiction between lawful norms and how they are set forth".[43] She cites Zincone who even says that "[this] discrepancy [...] between laws and their administrative implementation is typical of the Italian political system."[44]

The general awareness about the refugees' problems must have started in Italy during the eighties of the 20th century. Currle wrote that between 1984 and 1989 about 700,000 to 800,000 migrants had been coming to Italy and a bigger part of them had not been receiving residence permission. "The need for migration and integration policies got clear, because the immigrants of those days were coming to a country with almost no existing laws relevant to migration."[45] "It was not before mid-1980s that the country stated to realise that an immigration policy needed to be formulated […]", as Puggioni writes.[46]

Seen from a historical view point the first experiment of an immigration law apparatus was made in Italy 1931 according to Ventrella McCreight who has written about the laws from a martial law perspective. Immigration was detected as a problem of the public safety in those days. Above that you must be repelled as a foreigner if you were facing prison of one year or more or did attack the Italian state. According to decree 50/1948 from 1948 everybody who did not tell the identity of a foreigner or a stateless within 24 hours to the security institutions was to punish. The law 152 from 1975 said that migrants could be repelled who cannot show evidence for their financial safety in terms of their daily needs.[47]

The therefore following laws from the eighties onwards are described by Puggioni[48] as real beginning of immigration laws in Italy and Currle did not write about laws from the time before 1986. The decree 416/89 from 1989 defined the status of refugees for the first time. It became converted to the law 39/90 (Legge Martelli) in 1990 which meant to be a turn in

---

[42] see Currle 2004, pp.281f.
[43] see the same, p.285
[44] Zincone 1995, p.139, here according to Currle 2004, p.285
[45] Currle 2004, p.282
[46] Puggioni 2006, p.173
[47] Ventrella McCreight 2006, pp.143f.
[48] see Puggioni 2006, p.173

terms of laws about asylum seekers. I am going to write more about that in the following chapters. With the decree 489 from 1995 the law about the substantial evidence to be shown by a migrant about his or her ability to make a living in Italy was withdrawn. Because of that Ventrella McCreight wrote that the social aspect was been emphasized more than the one of martial law until then. According to her the martial law aspect began to become a priority with decree 286 from 1998. The supreme court (Corte di Cassazione) had called decree 286/1998 oppressive compared to the laws existing before that decree.[49]

Decree 286 is valid for non EU citizens and stateless people. It contains the punishment for persons who allow non EU citizens to come to Italy with up to three years in prison and the payment of up to 30 million Lire. If the violation of the law took place by three or more people or with the goal to profit financially and if five or more people are smuggled the decree foresees a prison penalty between four and twelve years and the payment of 30 million Lire per smuggled person. Police and jurisdictional collaboration on EU level had not been foreseen explicitly in decree 286.

With the beginning of an according collaboration on EU level from 2002 onwards a modification of the order of 1998 became necessary. The law 189/2002 (Legge Bossi-Fini) said that with bilateral relations with non-EU-countries of a kind which has nothing to do with human beings cooperation of those states with the prevention of illegal migration routes and criminal activities is appreciated. Such the law from 2002 seems clearer in regards to international relations as the law from 1998. Furthermore the Bossi-Fini-Law emphasizes on readmission agreements. After the law was in enforcement 39 contracts with non-EU-countries have been made. Hereby the law followed initiatives of the EU Commission about readmission agreements.[50] Ventrella McCreight cites Peers here and follows his statements as he says "[…] that EU policies on immigration, including readmission agreements, are '... unbalanced, inhumane, and internally contradictory'"[51]. Third the law would emphasize the cooperation between EU member states, because it explicitly cites the Schengen Aquis. With further explanations about the Bossi-Fini-Law Ventrella McCreight makes clear that it is a good fundament for the enforcement of EU laws and the additional protocol about human trafficking of the United Nations Convention against Trans-national Organized Crime (UNTOC).[52]

---

[49] Ventrella McCreight 2006, pp.144f.
[50] the same, pp.146ff.
[51] Peers: "Readmission Agreements and EC External Migration Law", from: "Statewatch" Nr. 17, from the internet, filed under: http://www.statewatch.org/news/2003/may/readmission.pdf, 18/03/2005, here according to Ventrella McCreight 2006, p.148
[52] see Ventrella McCreight 2006, pp.148ff.

Currle writes about the laws situation according to different topics for example "immigration and stay" and "asylum and refuge". Under "immigration and stay" Currle writes about the Bossi-Fini-Law from above from 2002 that since then non EU citizens had been allowed to come to Italy only if they already have had a valid work permit. It was possible before to apply for a residence permit within the first eight days of one's stay at the Questura (police head quarter).

With the Legge Martelli from 1990 it had been introduced for the first time that each year in October the number of immigrants from non EU countries according to the need for workers of the next year was published per decree. With the planning of the quote following criteria played a role: The development of the gross economic situation, the development on the job market and the capacity of the social system and the system of universities to admit new immigrants. Furthermore jobless immigrants who are in Italy already had a right to be accepted first while looking for a job.

Until 1997 the quote was always zero. Whereas family reunions, immigration by getting asylum and the immigration of refugees because of wars and civil wars from Yugoslavia and Somalia as well as refugees from Albania have not been counted. For 1998 Currle gives the quota of 27,000 and for 1999 of 58,000 regarding workers from non EU countries. The quota for 2000 of 63,000 was fulfilled in June already. Through the need for workers in the low income sector the quota was seen as a means to give the industry enough workers working for a low salary. Especially in the industry cities of northern Italy there are not enough Italian workers who want to work for such a low summary any longer. The development of the population would also play a stronger role more and more as Italy has got one of the lowest birth rates worldwide as well as a high percentage of elderly people. Since 2002 the calculation and proclamation of the quota lies in the hands of the government, but it is not obligatory any longer, but facultative.[53]

It is particularly interesting what Currle writes under point "Asylum and Fleeing". In article 10 of the Italian constitution the right of finding asylum is written down, but only under specification by laws. According to the law every foreigner has the right to get asylum who is hindered from using one's democratic rights in his or her own country. A clear and specific law does not exist until now, but the practical processes show a normative meaning. Italy would follow the regulations of the Genf refugees' convention (For the observing reader it is clear that I have already cited other sources like Milborn who are complaining about the opposite). Until 1990 the refugees' convention had been enforced only with refugees from European countries. Until 1990 the refugees' law had been enforced only with refugees from

[53] Currle 2004, pp.286f.

Europe. The Martelli-Law from 1990 disestablished that geographical limitation again.

Between 1990 and 2002 the proving of the asylum applications took place at the central commission for the admission to the refugee status (Commissione centrale per il riconoscimento dello stato di rifugiato). The Bossi-Fini-Law modified also the asylum sector and replaced the central commission through local ones.

During their process asylum seekers would not be allowed to work in Italy. Public financial aid exists only for a very short period. Admitted refugees would not get regular subsidies or aid for a rehabilitation within society, but are seen as possessing the same rights as Italian citizens concerning the right for education, access to the health service and public financial relief.[54]

Gleitze / Schultz who made interviews with refugees relief organisations, scientists and camp inhabitants wrote about the law from 2002 that it is characterized through four "massive restrictions". Firstly the information centres for asylum seekers (Centri d'identificazione, CDI) had been built. Second there is a regular asylum procedure and there are also simplified procedures in a CDI. And third the territorial commissions (the "local commissions" from above) in Gorizia, Milano, Rome, Foggia, Syracuse, Trapani and Crotone had been introduced. Fourth an appellation can only get priority if the prefect admits it in certain causes[55]. In chapter 3.2 I am going to write in more detail about the practical consequences of the Italian laws.

2001 there was the attempt to introduce an asylum law, but before it had been decided about it, the president put elections that took place to elect a new parliament and the right wing politician Silvio Berlusconi was the winner in the elections.[56]

In the following chapter the practical handling of laws concerning the boat people is explained. Hereby the above mentioned gap between law and enforcement exists.

3.2 In Practice: Calculated Chaos?

Gleitze / Schultz complain that asylum applications either shown at police stations at the border or in the police headquarters (Questure) often are not accepted or given ad acta for a long period of time. If an application is accepted it gets forwarded to the territorial respectively local commission. Only a few of those commissions are functioning in practice so that asylum seekers have to cope with long travels to the commissions, partly through the whole of Italy. The scheduled time for the process would not be fulfilled according to their

---

[54] see Currle 2004, p.288

[55] see Gleitze / Schultz 2006, p.6, from the internet as mentioned above

[56] see the same

experiences according to Gleitze / Schultz.[57] Puggioni also criticizes the police as she writes about the not existing handing on of information about the asylum process and an arrogant behaviour which can be seen in general concerning the police forces. Furthermore she reports that deportation decrees are enforced automatically even if the refugees came from countries that are well known for human rights violations. Puggioni implicitly assumes structural racism[58]. Because she reports that there is a trend to let Kurds immigrate and give them a chance for a fair asylum process whereas Albanians, Roma and Maghrebian (People from northern Africa) were repelled quickly and Chinese people instantly at the borders.[59]

For the accommodation of refugees and asylum seekers the Sistema nazionale di protezione (National asylum programme) is valid. Whereas the old PNA project (Programma Nazionale Asilo, National Asylum Programme) offered about 2,000 places in communal managed flats and smaller residences, the new sistema nazionale di protezione foresees in the first line the CDI which means forced integration in a dormitory.[60] From the maximum up to between seven and eleven CDIs "[...] it seems as if only the ones in Crotone (Kalabria) and in Siracusa (Sicily) are existing." The commissions would be integrated in the CDIs directly and the hearings would take place there.[61] The CDIs would have the same restrictions as the prisons for asylum seekers. From outside there is no possibility to enter the buildings which makes the counselling and taking care through independent lawyers and organisations very difficult. The placing into a CDI means for an asylum seeker that he or she is a clandestine refugee with an illegal status and only the simplified process is valid for him/her. Because of the new law there are almost only simplified processes where the right to appeal is restricted. Besides the CDI there are further refugee camps that I am going to describe more detailed below.

The commissions enforce residence permits in cases of acceptance and for the subsidiary safety of the asylum seekers according to Gleitze / Schultz[62]. Persons with a right to get subsidiary shelter are people without any citizenship and citizens of non EU countries who do not fulfil the preconditions for the acceptance as refugee, but can show evident reasons for their individual suggestion of a danger for them (Council Directive 2004/83/EC) as "Migration und Bevölkerung" reports.[63]

---

[57] see the same, pp.6f.
[58] Puggioni 2006, p.179
[59] Consorzio Italiano di Solidarietà: "Dossier Nausicaa. Primo Quadro sulla Tutela del Diritto d'Asilo in Italia", December 2000, unpublished paper, p.22 und p.25, here according to Puggioni 2006
[60] see Gleitze / Schultz 2006, p.7, from the internet as mentioned above
[61] see the same, p.8
[62] see the same, p.7
[63] see „EU: Initiativen in den Bereichen Migration und Grenzschutz", from „Migration und Bevölkerung" 5/2007, exclusively in the internet, filed under: http://www.migration-

Concerning the personnel of the commissions they contain a member of the prefect institution, the questura, the Carabinieri, the ministry of the interior and the UNHCR. The law expert Fulvio Vassallo Paleologo of the University of Palermo called all members not independent. Some of the interview partners of Gleitze / Schultz even said about the UNHCR that it is responsible that the hearings are still handled "that strict". Because of a lack of funding the representative of the UNHCR was not able to be present in Italy during the travel of Gleitze / Schultz (reported about in April 2006).

As the biggest problems during the hearings Gleitze / Schultz see:

- "No guaranteed presence of an interpreter with whom the refugee is able to communicate (often there are only English or French speaking interpreters involved);
- Mean time for the hearings is 5 - 10 minutes per person;
- Mostly there are no lawyers present, because the refugees often do not get any information about their trial and their right to order a lawyer;
- Presence of personnel of the embassy of the country of origin."

The fundament for the hearing by the commission is the first interview by the police.[64] Giorgio Bisagna, lawyer and coordinator for the CIR (Italian refugees council) in Sicily says that there is a problem not so much concerning the right of a fair process, but concerning the facts that the refugees are under detention in prisons without possibilities to enter and on the other hand that they themselves have to pay for their lawyers who could possibly work for them. "The most scandalous with that system is there is now effective help during the trials."[65] About repelled asylum seekers I am going to write something again at the end of this chapter.

---

info.de/migration_und_bevoelkerung/artikel/070508.htm

[64] see Gleitze / Schultz, pp.19f., from the internet as mentioned above

[65] Giorgio Bisagna in his interview at 27/4/2006 in Palermo, here according to Gleitze / Schultz 2006, p.21, from the internet as mentioned above

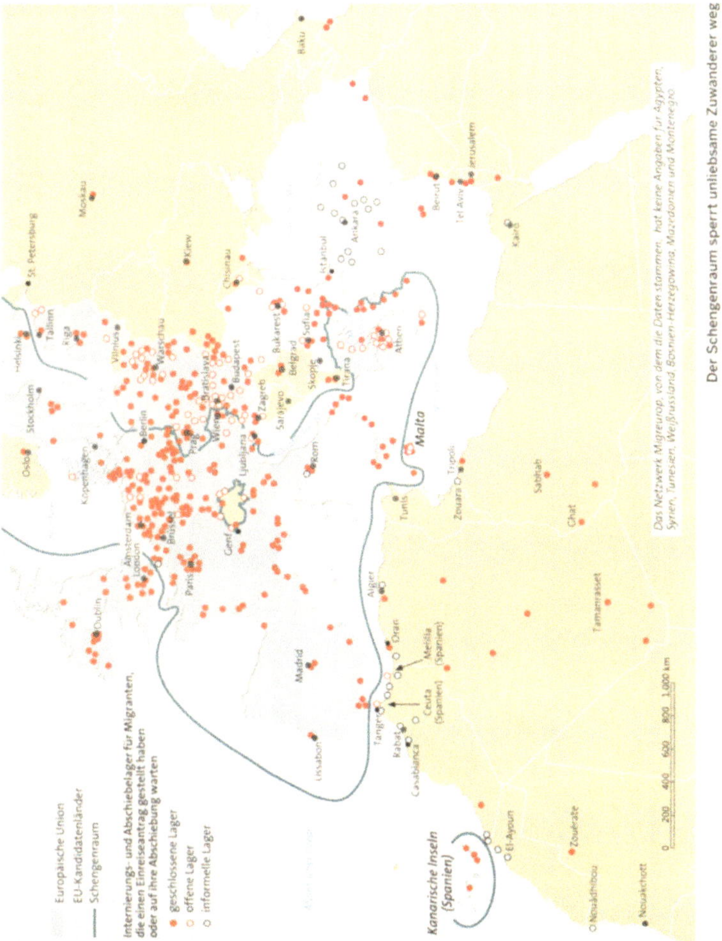

Picture 4: The Schengen Territory places unwanted migrants under detention (red dots mark closed camps) - „Le Monde diplomatique" / taz Verlags- und Vertriebs GmbH 2006, p. 51

The necessary prolongation of the residence permit by the commission needs a lot of time and can only be applied for at the Questura (police headquarter) which was in charge during the crossing of the border which again can mean a long journey. "Asylum seekers with denied asylum either come under detention or are casted adrift with a so called „foglio di via" (emigration/deportation order) and have to leave Italy within five days." As soon as they get the "foglio di via" there is the possibility to appeal against it. Many refugees get the order in advance immediately after their arrival before they apply for asylum. Whereas, because of the new law there is no chance to stop the enforcement of the order immediately! It is said that then during the more simplified procedure it was possible to get a new hearing within five days. This has to be before the commission which has denied the application. Then it would be unrealistic that this commission withdraws its own decision. The second possibility is that the asylum seeker with the help of a lawyer who he/she has to pay for, hands on an application for immediate enforcement to the prefect (head of the police?) within 15 days after the decision of the commission. However, the application has to be delivered to the prefect within those 15 days which is also unrealistic. This way the asylum seekers find themselves in a dead end.

After the detention in a CDI churches, welfare associations and other organisations take care of the refugees. But they cannot accept all of them. An official financial support of 45 days by the state would often not be paid for the refugees.[66]

Overall there would be a big number of regular refugee camps. The centres that welcome the refugees are the so called Centri di Prima Accoglienza (CPA). Asylum seekers are forwarded to the CDIs. Refused asylum seekers can be deported to a prison (Centro di Permanenza Temporanea e Assistenza, CPT) since 1998 (law of the middle-left coalition „l'Olivo"). The maximum duration of the detention would have been prolonged from 30 to 60 days with the Bossi-Fini-Law. Also refugees without a status of residency who have applied for asylum, but already have got a "foglio di via" come into a CPT. CPTs are run by clerical organisations and the Red Cross. The watchmen are from the government. Only in rare cases the asylum seekers get an entry to centres of the second welcoming (communal institutions, Centro di Seconda Accoglienza).[67] The present programme with fewer than 2,000 places serves only 6.76 percent of the really needed number of places[68]. Partly there are combined centres of CPA, CDI and/or CPT.

Gleitze/Schultz report further that after their arrival at Lampedusa the refugees should be brought to another camp at the continental part of Italy for a longer stay within 48 hours according to the laws. The two authors hereby cite Vassallo Paleologo:

> "Lampedusa is also seen as a CPA regarding the present circumstances if a person is hold under detention. Because of emergency more than 48 hours can pass in a CPA – here the dividing of holding under detention and welcoming is very difficult, often even not existing."[69]

A statement of the lawyer Francesco Vizza about the combined centre Sant'Anna in Crotone shows the chaotic circumstances there very clearly according to Gleitze/Schultz:

---

[66] Gleitze / Schultz 2006, p.7, from the internet as mentioned above
[67] the same, pp.8f.
[68] ICS (Consorzio Italiano di Solidarietà): La protezione negata (2005), p.137, here according to Gleitze / Schultz 2006, p.8, from the internet as mentioned above
[69] Fulvio Vassallo Paleologo in his interview at the 01/05/2006 in Palermo, here according to Gleitze / Schultz 2006, p.9, from the internet as mentioned above

> "Well, the different sort of detention is the formal aspect. Sant'Anna is a big territory which is divided into building A, B, C and D. The Zone A is a CPA, but it depends how many arrive: If for example there are 500 arriving at once, zone A of course is not enough. And also zone B becomes a CPA. Anyway, it is all similar."[70]

Gleitze / Schultz had an interview with MSF (Medici senza Frontieri, Doctors without Borders) which is active at Sicily since 2003. They manage the first aid of the arrived refugees as well as the observation of the landing and arrival as whole and have made it their business. MSF is called to each of the locations of a refugees' transit stay by the health authority (ASL) during the landing of the refugees. From there the refugees are transferred to the different camps. An MSF representative reported about such a location, an aeroplane hangar, in an interview:

> "[...] sometimes 200 arrive at once. The hangar in Pozzallo is very big. It is big enough for so many people. [...] That is a location without any juridical status which does not exist. [...] Once they put 170 people under detention at once in a school in two or three little rooms."[71]

Concerning refused asylum seekers the scientific report about asylum "La Protezione Negata" (The denial of shelter) describes the following process as regular for Italy:

> "Commissions follow standardised formulations, decide according to ethnicity, without any individual proof of facts, go for alleged contradictions which on the other hand could hardly be evident as an outcome of the `long` first interview after the arrival and the even `deeper digging` hearing. Often the foreign affairs have an influence on the decisions according to the authors. After the denial and the enforcing of the foglio di via there is only the counter accusation against it left. Because this has no effect that hinders the enforcement immediately they can apply for this effect at the police head quarters. During this period there is also a need for a residence permit because the refugees would be staying in the country without a legal status during the trial. But this permit can only be applied for from the same judge who decides about the process because of the accusation for non-enforcement of the foglio di via. This means de facto that most of the refugees do not possess a residence permit, but only a copy of the reason for their counter accusation. Because this does not count as a residence permit many get into the danger of being controlled and repelled to their country of origin."[72]

The lawyer Vincenzo Medici from Crotone reports about one of his clients whose cause would be "a good example for the Bossi-Fini-System". During his arrival in Italy he had got the foglio di via. His hearing before the commission was denied. After this he stayed under detention for 60 days later he got the usual order to leave the country within five days. The client was checked by the authorities and came under detention for 60 days again, came free again and so on.[73] "As soon as an asylum seeker gets a denial he or she stays illegally. There is no tolerated residence in Italy", as Gleitze / Schultz cite Karen Basile from the counselling

---

[70] Francesco Vizza in his interview at the 21/04/2006 in Crotone, here according to Gleitze / Schultz 2006, p.11, from the internet as mentioned above

[71] MSF, 25/04/2006, here according to Gleitze / Schultz 2006, p.9, from the internet as mentioned above

[72] ICS (Consorzio Italiano della Solidarietà): "La protezione negata" (2005), p.73f., here according to Gleitze / Schultz, p.23, from the internet as mentioned above

[73] Vincenzo Medici in his interview at the 21/04/2006 in Crotone, here according to Gleitze / Schultz 2006, p.23, from the internet as mentioned above

centre of the UIS/UIL Union in Palermo.[74]

Unaccompanied under aged refugees build a special case. Often they are "made older" by the authorities. And they get under detention together with other refugees in the camps. Because there is often hardly any suitable place for them they often move around in the country and are a potential victim of exploitation and sexual harassment. The report of Gleitze / Schultz contains more information about that.[75]

At the end of the short version of their report Gleitze / Schultz report that daily the Italian practice of the denial of rights for refugees with its special mixture between tyranny and political favoured chaos leads the international regulations about refugees ad absurdum. "The EU member countries, partner in the process of the European standardizing of the asylum laws stay silent about it. It is the silence of collaboration."[76]

[74] Karen Basile in her interview in Palermo at the 27/04/2006, here according to Gleitze / Schultz 2006, p.23, from the internet as mentioned above

[75] see Gleitze / Schultz 2006, pp.29-32, from the internet as mentioned above

[76] Gleitze / Schultz: „In den Zonen der Rechtlosigkeit", from the internet: http://www.proasyl.de/de/italien/italiens-asylsystem/index.html, 20/12/2007

4. The EU Policy

After a catastrophe with refugees involved in October 2003, when Italian fishermen rescued 14 refugees from a boat with about 100 people, Prime Minister Berlusconi urged from the European Parliament the solidarity of the rest of the member countries: “We cannot shut our eyes in front of this tragedy. The rich, Christian Europe has to think how it welcomes those who search for a good future here.” Besides the Italian minister of the interior Giuseppe Pisanu spoke about common border patrols, about common quotas of refugees as well as for pushing development cooperation with the countries of origin.[77] Those statements sound like honest goals and an ambitioned policy which goes to the roots of the problem. The following shall give a short overview over the EU migration policy and especially the EU border control agency Frontex.

4.1 Adjustment of the EU Migration Laws

Bade describes how the European migration policy developed to a defence policy for the control of the borders of the EU hostile against the refugees. Since the late 1980s there would have been strengthened “a position of defence and a recipe of migration policy as security policy“. Bade begins with the “asylum aquis” in Germany from 1992. Hereby the clause of the first welcoming country and the safe country doctrine was introduced to the German elementary right of asylum. “Since then no one had got the chance to get asylum in Germany any longer who comes from countries `without prosecution` - whereas Germany has introduced this regulation for all neighbouring countries without their authorization.”

The German restrictions had coerced other European countries to introduce restrictive measures, too. Through that they wanted to limit the immigration to their own territories. Bade talks here about a “chain reaction in a negative coalition of defence which grows together”. But this alone had hardly lead to supra national concepts of defence, according to Bade. Those would be founding on the integration of countries within the EU which lead to an enforcement of the outside borders of the EU, “because of the so called rising `vulnerability` through migration as a danger”.

The way to the coherent asylum and immigration laws in the EU would still be a far one as Bade writes in 2000. He identifies at least four great groups of typical forms of immigration. They are the result of various regulations, restrictions and denials that are valid for third countries since 1990:

---

[77] „Italien: Hilfe bei Grenzsicherung gefordert“, from: „Migration und Bevölkerung“ 9/2003, from the internet, filed under: http://www.migration-info.de/migration_und_bevoelkerung/artikel/030906.htm

1) Families' reunions
2) Privileged relations of migration
3) Migration because of job seeking (Elites' migration and migration of underprivileged people)
4) Migration of refugees and deported persons

The families' reunions become more at the cost of other groups, because they only include those who already have tight connections with their new home country, want to get them (`marriage migration`) or try to create those for their children through giving birth to them (`birth migration`). As privileged migration connections Bade describes post colonial immigration or also the immigration of "Aussiedlers" (people with German roots emigrating to Germany from the former Soviet Union) and at a lower level the immigration of Jewish people from the GUS (German abbreviation for: Gemeinschaft unabhängiger Staaten, which means "Community of independent countries"). The functional elites that are mobile anyway and seen as politically not dangerous respectively irrelevant for the migration control are belonging to the field of the job seekers' migration. At the basement of this pyramid there are opposite the members of a migrants' class with a mostly only seasonal allowance of immigration to work in the country which leads into sectors and job relationships that people who are born to this country do not any longer want to have for themselves.

Concerning the fourth form, the one of the refugees and deported persons, Bade writes: "The fences of the `fortress Europe` got higher. The search to irregular entries and the fight against this became very important topics within the European discussion about migration."[78] I am not going to write about the political frame works of the EU level in detail here, favouring more the following topics who found an echo of criticism from refugee and human rights organisations.

4.2 The Frontex Agency

The EU commissioner for jurisdiction and the interior Franco Frattini talked about a "two-pronged" strategy fighting illegal immigration from Africa by sea in his speech at the meeting of the ministers of foreign affairs and the ones of the interior of Mediterranean countries and the EU in 2006. Firstly there should be started a series of operational and political measures to fight illegal immigration, make sure the shelter for refugees and a strengthened enforcement of the border control and surveillance at the maritime borders. Second there should be a political framework to be developed together with the countries of origin and the ones of migrants' transit in Africa.

---

[78] see Bade 2000, pp.16-19

After that Frattini talks about the Frontex agency when he talks about the operational and political measures that can be done on the short term. Frontex would have done a great job through coordinating the border control near the Canary Islands and at the central Mediterranean Sea. Frattini gave the order to the member countries to let the budget of Frontex rise during the following years.[79] The budget rose from 19.2 million to 35.2 million Euros between 2006 and 2007[80], and was even doubled to 70 million Euros for the following year 2008[81]! Whereas the budget of the European Migration Network (EMN) in charge of the coordinating of collecting and analysing data about migration and asylum stays at only three million Euros in 2006.[82]

However, what is this Frontex agency now? Frontex is the "European Agency For The Management Of Operational Cooperation At The External Borders Of The Member States Of The European Union (FRONTEX)", which means that it is the EU border control agency.

> "Besides other things Frontex is in charge of coordinating patrols on sea and by aeroplanes to catch potential immigrants in their boats at the coasts. In cooperation with concerned EU member countries Frontex also prepares the deportation of caught migrants."[83]

Frontex was founded with the EU council regulation 2007/2004 from 26th October 2004 and resides in Worsow since the 26th of April 2005. And in 2005 it also begun its work.[84] In February 2007 „Migration und Bevölkerung" reports about the up to then biggest Frontex operations near the Canary Islands including the cooperation of seven EU respectively Schengen countries between July and December 2006. Within the framework of "Hera I" there would have been registered 19,000 persons who immigrated clandestine. Through "Hera II", the second module of this operation, the refugees should already be hindered from leaving the African coast. Frontex was being patrolling with ships and aeroplanes at the coasts of Senegal, Mauretania, the Cape Verdeans and the Canary Islands. All in all more than 3,500 migrants would have been hindered from going to European territory.[85]

Concerning the part of the sea between Sicily and Africa the organisation "Fortress Europe" has reported about the Frontex operation "Nautilus I" (5th to 15th of October 2006) and "Nautilus II" (25th of June to 27th of July and 10th of September to 14th of October 2007). Italy, Malta, France, Greece, Germany, Portugal and Spain were cooperating in those operations.

[79] see conclusion of the speech of Franco Frattini at the 29/09/2006 in Madrid, from the internet, filed under: http://ec.europa.eu/commission_barroso/frattini/archive/speech_29_09_06_en.pdf
[80] see Frontex / Finance, from the internet, filed under: http://www.frontex.europa.eu/finance/
[81] see Fortress Europe 2007, p.3, from the internet as mentioned above
[82] see Commission of the European Communities 2005, p.5, from the Internet, filed under: http://ec.europa.eu/justice_home/doc_centre/immigration/docs/com_2005_606_de.pdf
[83] „Migration und Bevölkerung" 1/2007, p.4, from the internet, filed under: http://www.migration-info.de/migration_und_bevoelkerung/archiv/ausgaben/ausgabe0701.pdf
[84] vgl. Frontex / Legal basis, from the internet, filed under: http://www.frontex.europa.eu/legal_basis/
[85] „Migration und Bevölkerung" 1/2007, p.4, from the internet as mentioned above

They were planning to introduce permanent patrols under the auspices of Frontex by aeroplane and by sea at Libyan parts of the sea as Frattini already had been saying. The deportation of refugees to Libya who were caught at sea would be the clear goal.[86] A prototype of the Frontex operations was the Neptune project at the central and eastern Mediterranean Sea during which Italy, Germany, Greece, Malta, Cyprus, France and the United Kingdom were taking part from September 2003 onwards through common forces with patrols by aeroplanes and at sea.[87]

In May 2007 the European Patrols Network (EPN) was founded by Frontex. The members of the network are the Mediterranean countries France, Greece, Italy, Malta, Portugal, Slovenia, Spain and Cyprus. After a phase of coordinating the patrols by Frontex there shall be extended coordinating centres that are in charge for certain coastal zones. Interested third countries shall be included into the network. Libya has already been mentioned above as first partner in 2008.

The European Parliament decided for a regulation about rapid task forces in April 2007. They can be ordered by member countries in exceptional and emergency cases in other words a high number of immigrations of clandestine persons at once.[88] In August 2006 you have already been able to read about task forces of 250 to 300 border control officers and experts that are said to be able to act within ten days in „Migration und Bevölkerung".[89]The task force regulation would be a deal between the European Parliament, the EU Commission and the European Council according to a report in 2007. The European Parliament would have completed the proposal of the Commission first of all through additional clauses regarding the respect of human rights as well as regarding to the necessary solidarity between the member countries.[90]

Regarding to a juridical report the human rights organisation Amnesty International and the already mentioned refugees' aid organisation Pro Asyl criticized that Frontex would not respect duties of the EU member countries in regarding to human rights and refugee laws through its defence against refugees. The criticism was published on the day of refugees in September 2007. Also outside of the territories of EU member countries – as well as at sea outside the 12 miles zone – the border officers would be bound to the refugees' rights and human rights. Refugees who got caught at sea would have the right to apply for asylum

[86] see Fortress Europe 2007, p.3, from the internet as mentioned above
[87] see Cuttitta 2004, p.133
[88] „EU: Initiativen in den Bereichen Migration und Grenzschutz", from: „Migration und Bevölkerung" 5/2007, exclusively in the internet, from the internet as mentioned above
[89] „Migration und Bevölkerung" 6/2006, p.3, from the internet, filed under: http://www.migration-info.de/migration_und_bevoelkerung/archiv/ausgaben/ausgabe0606.pdf
[90] „EU: Initiativen in den Bereichen Migration und Grenzschutz", from „Migration und Bevölkerung" 5/2007, exclusively in the internet from the internet as mentioned above

regarding to this report. It would also not be allowed to repel them if there is a threat of harassing and mistreatment for them. "To simply hinder refugees of staying travelling or to accompany them back to the coast is forbidden according to this report. They have the right to be brought to the next safe haven", said Günter Burkhardt, CEO of Pro Asyl.[91]

Certain questionings from the German Bundestag (The federal parliament) about the turning away of boats or the practice during the catching of ships would not be answered precisely, but unclear by the German government according to Burkhardt in his statement at the press conference on the 27th of September 2007 in Berlin.

Concerning parliamentarian questionings about Frontex it seems as if also the council of the European Union is hesitating to answer them. To numerous and detailed questions by the German member of the Bundestag Angelika Beer (green party) about the role of Frontex concerning charter flights for the deportation of refugees the council answered her as follows: "The council would like to inform the MP that the special information and facts she asked for are not available." Following to this answer the council added two paragraphs about the juridical position of Frontex and a few general statements and that was all what the council answered.[92]

The unclear answering of parliamentarian questionings through the German government would have been one reason for the foundation Pro Asyl, Amnesty International and the Forum Menschenrechte (forum human rights) to order the juridical report „Menschen- und flüchtlingsrechtliche Anforderungen an Maßnahmen der Grenzkontrolle auf See" (human rights' and refugees rights' regulations of measures of the border control at sea). As an example Burkhardt talked about a case at the 10th of September 2007 (that was during "Nautilus II") and the measures of the Italian coast guards on this day. Tunisian fishermen had saved 44 refugees coming from Sudan, Somalia, Eritrea and Morocco from a damaged rubber boat. The coast guards are said to have tried to go side to side with the boat of the fishermen to turn it away and to have given the fishermen the order to go back to Tunisia. The fishermen had already been threatened with detention at sea in case they would bring the refugees to Lampedusa. In the following trial against the fishermen it got obvious that the coast guards were being acting according to and fulfilling of an order. Pro Asyl was getting to know the details of this case from the above mentioned Fulvio Vassallo Paleologo from the University of Palermo. Whereas Burkhardt says that Frontex would act using finer methods, those would

---

[91] see Pressemitteilung Amnesty International, Sektion der Bundesrepublik Deutschland, from the internet, filed under:
http://www.proasyl.de/fileadmin/proasyl/fm_redakteure/Europa/Frontex/Fluechtlingstag_27Sep07__2_.pdf
[92] see Council of the European Union: Vorentwurf einer Antwort auf die schriftliche Anfrage E-3720/07 von Angelika Beer, from the internet, filed under:
http://www.proasyl.de/fileadmin/proasyl/fm_redakteure/Newsletter_Anhaenge/127/PE-QFRONTEX.doc

also be clearly against the laws.[93]

The report ordered by the refugees' aid and human rights organisations has been written by Andreas Fischer-Lescano and Tillmann Löhr from the European Center for Constitutional and Human Rights (ECCHR). In their summary Fischer-Lescano / Löhr write besides else things that people who are in need of shelter and migrants who get into an emergency situation at sea have to be treated according to the regulations of the humanitarian jurisdiction at sea. It would be forbidden to bring saved persons to third countries where there is no substantial safety guaranteed for them (unlawful refoulement). The ones who seek for shelter would have the right to be brought to the next save haven on European territory.[94]

According to the Genf refugees convention, article 33 (the first paragraph) the principle of non-refoulement which is "[called] the fundament of international laws about the safety of refugees" by the UNHCR says as follows:

> "No one of the countries that have signed this convention will in any way repel a refugee and let him or her go back across borders to territories where their life or freedom could be at risk because of reasons of their race, religion, citizenship, his or her membership in a certain social group or his or her political opinion."[95]

Fischer-Lescano / Löhr write in their conclusion that within the framework of the European surveillance und emergency measures under collaboration with third countries there exists the duty for the border control officers to bring people who seek for shelter and migrants to a safe location where the laws about the refoulement are particularly respected: "This is not guaranteed in the African transit countries, therefore the transport of people to the EU member countries is a must."[96] The Frontex mission at sea in the north-west African coastal zones and the ones of the Canary Islands in 2006 with the defined goal "to watch out for boats with illegal migrants on board, to catch them and to accompany them back to the African coast" (according to "Migration und Bevölkerung" 6/2006)[97] was against fundamental international laws, because of that written above.

In the last chapter of the main part of this study I am going to write also about the situation concerning Libya the already mentioned partner of the EU and Italy in affairs of illegal migration.

---

[93] see „Auszug aus dem Statement zum Pressegespräch am 27. September 2007 in Berlin", from the internet, filed under: http://www.proasyl.de/fileadmin/proasyl/fm_redakteure/Europa/Frontex/Statement_Presse-gespraech_am_27_September__2_.pdf

[94] Fischer-Lescano / Löhr 2007, p.30, here according to Pro Asyl, from the internet, filed under: http://www.proasyl.de/fileadmin/proasyl/fm_redakteure/Europa/Frontex/Gutachten_Geltung_MRe_auf_See_Sept._2007.pdf

[95] see UNHCR 2007, p.4, here according to Pro Asyl, from the internet, filed under: http://www.proasyl.de/fileadmin/proasyl/fm_redakteure/Newsletter_Anhaenge/123/Diplomatische_Zusicherungen.pdf

[96] see Fischer-Lescano / Löhr 2007, p.30, here according to Pro Asyl, from the internet as mentioned above

[97] „Migration und Bevölkerung" 6/2006, p.3, from the internet as mentioned above

4.3 The case of Libya: Criminal Collaboration with the EU and Italy

"Third countries like Libya, Morocco, Mauretania, the Ukraine, Turkey and so on get the function as gatekeepers at the gates to the `Fortress Europe` in a cynical delegating of tasks", says Karl Kopp from Pro Asyl in his article for the day of refugees in 2007. He writes about an example of severe human rights violations in those third countries. In Morocco over 400 Sub-Saharan refugees had been caught and abandoned at the Algerian border between Christmas 2006 and New Year 2007. There had been severe mistreatments by the Algerian and Moroccan security forces during which several women got raped and one pregnant woman lost her unborn child. "That human rights violation fits together in a row of brutal acts against shelter seeking people in Morocco – and Europe keeps being silent about this and is looking away."[98]

"The observatory on victims of migrations FORTRESS EUROPE" expresses an accusation in its report about the situation of transit migrants in Libya which sounds quite similar. In this report ("Escape from Tripoli") they write about crimes by the people who smuggle other people ("passeurs") as well as by the Libyan authorities against the migrants which are pointed out first hand by interviews: Mistreatment, harassments, irregular capture and detention under bad conditions without a trial, torture, physical and sexual violence, the deportation of refugees to their country of origin and deportations to the desert. Crimes the EU allegedly does not see, since it has allowed the refoulement of migrants to Libya through Frontex patrols.[99]

2004 the Italian government has enacted a bilateral treaty with Libya for the fight against illegal migration. The contents of that treaty never have been published despite questionings of MPs as Silja Klepp wrote in the brochure for the day of refugees in 2007.[100] According to Milborn the Italian Government has already signed a "secret treaty" with Libya about deportations and the building of camps in 2003.[101] Both write that an EU mission to Libya (in April 2005, according to Milborn) later detected that the treaty contained the delivery of technical equipment for the border control, furthermore according to Milborn woollen sheets, mattresses and 1,000 sacks for dead bodies. As Milborn writes, Italy furthermore paid for 60 flights to repel refugees from Libya to other African and Asian countries and has financed one camp in the north and two in the south, near the towns of Kufra and Sebha since 2003.

Libya has not signed the Geneva refugees' convention. The UNHCR is at the site in Tripoli,

[98] see Pro Asyl (Ed.): „Tag des Flüchtlings 2007", p.33, from the internet, filed under: http://www.proasyl.de/fileadmin/proasyl/fm_redakteure/Archiv/HTdF_Hefte/Webversion_Heft_komplett.pdf
[99] see Fortress Europe 2007, pp.3f., from the internet as mentioned above
[100] see Pro Asyl (Ed.): „Tag des Flüchtlings 2007", p.37, from the internet as mentioned above
[101] see Milborn 2006, pp.58f.

but is not recognised by the Libyan government, and because of this cannot work without restrictions there.[102] The country is partner at the unofficial 5+5 conferences together with the five EU countries Portugal, Spain, France, Italy and Malta and the four northern African countries Mauretania, Morocco, Algeria and Tunisia. The group of countries has been dealing with illegal migration movements in the Mediterranean area since 1990 already, according to the human rights expert Paolo Cuttitta from the University of Palermo.[103] I am going to write about that in the last chapter of the main part. I have not been able to find literature about the 5+5 conferences in the internet, which goes beyond general statements.

According to Fortress Europe they already have published a report about a "technical mission" of the EU to Libya in December 2004 which contained the first query about the circumstances under whom the refugees have to live there: "It reported the arbitrary arrest of foreigners, abuses, collective deportations and the failure to recognise the right to asylum."[104] The report of the EU mission had stated that there is absolutely no asylum system in Libya according to Klepp. Nevertheless the report would come to the "surprising conclusion" that the cooperation with Libya would be important and needs to be achieved.

The members of the Frontex-lead technical mission of the EU in 2007 cannot be called independent observers. According to their own statements the team consisted of four representatives of Frontex, one representative of the department for immigration of the Maltese police and one Italian officer who has got his working place in Tripoli, which gives the number of six members of the technical mission. The report shows clearly that it is not probable that there existed much interest in humanitarian aspects except of a superficial comment about the asylum laws (on page 9 of the report). Technological aspects of the border control and the cooperation of Libya with Frontex have been stressed constantly in the report.[105]

Over 22,000 refugees had come to Italy via Libya in 2006, writes Klepp. When in the recent past job seeking migrants have been invited to Libya with an open minded immigration policy, today extreme racism exists broadly within the society of Libya against coloured inhabitants. Immigrants who have worked in Libya for a long time already cannot feel safe any longer, too. Examples for the situation on the streets of Tripoli are given also in the report of Fortress Europe. A youth from Eritrea is cited there with the following words:

---

[102] see European Commission 2004, p.6, from the internet, filed under: http://www.statewatch.org/news/2005/may/eu-report-libya-ill-imm.pdf, seemingly not contained in official EU websites

[103] see Cuttitta 2004, pp.131f.

[104] see Fortress Europe 2007, p.4, from the internet as mentioned above

[105] see Frontex 2007, from the internet, filed under: http://www.proasyl.de/fileadmin/proasyl/fm_redakteure/Newsletter_Anhaenge/130/eu-libya-frontex-report.pdf

"If they attack you and you call the police, you're the one who is arrested. Against us, they are twice racist. We are black and Christian. I was walking with a friend, Goitom, in Tripoli, when a group of Libyan youths stopped us. They wanted to know what our names were. So we presented ourselves with Muslim names. I said I was called Muhammad and Goitom introduced himself as Mustafa. Then they asked us if we were Muslim. We said that we were, luckily we spoke Arabic well. Not convinced yet, they asked us to recite the Fatiha sura from Koran. I knew it, but Goitom remained silent. They let me go. I ran to hide behind a small wall. I heard shouting, they were beating him, they wanted money. Goitom had 900 dollars on him, they were meant for the journey. He tried to defend himself. But they stabbed him and took everything away from him. When they had left, he was already dead, in a pool of blood."

A man from Nigeria reported that he witnessed two murders of young migrants in police stations.[106] Caught by the police refugees are repelled in aeroplanes. They are also been brought to the dessert and abandoned there.

The deportations from Italy to Libya have been stopped at the moment, since the change of the regime. Nevertheless, the further way of the EU cooperation with Libya seems to be predictable regarding to Klepp. "Furthered by the leading role of Italy the cooperation of the EU with the Gaddafi-regime develops faster and faster."

4.4 Development Cooperation: Blackmailing or the Solution?

The above mentioned article of Cuttitta describes the 5+5 group as the following:

"The 5+5 group wants to further the legal immigration and the integration of the migrants, do something against the emigration from the countries of origin by providing development cooperation and against the illegal migration by the signing of take back agreements between the EU respectively its member countries and their Mediterranean partners."[107]

Cuttitta describes development cooperation as a political means of forcing other countries to behave in a certain way (here used by Italy) against the African third countries. Immigration quotes get lifted upwards to achieve the signing of take back agreements and development cooperation is offered if the concerned countries are willing to cooperate in affairs of illegal migration. The concerned countries do not have the position to make a good deal with their counterparts so that the director of the Ghanaian immigration authority was upset and said: "To claim that we either have to take back our migrants, or the development cooperation will be undergoing new discussions is nearly blackmailing."[108]

But development cooperation can also be seen as a way to go to the roots of the problem, as Kevin John O'Connell makes clear in his article referring to the incidents in Ceuta and

[106] see Fortress Europe 2007, p.13 and p.15, from the internet as mentioned above

[107] see Conclusions of the Euro-Mediterranean Mid-Term Meeting of Foreign Ministers, Crete, 26-27 May, 2003; Conclusions of the Euro-Mediterranean Conference of Ministers of Foreign Affairs, Naples, 2-3 December, 2003; Relevé des conclusions de la Présidence and Proposition de mise en œuvre de la Déclaration de Tunis la 2ème Conférence Ministérielle sur la Migration en Méditerranée Occidentale (Rabat, les 22 et 23 Octobre 2003), all here according to Cuttitta 2004, p.132

[108] In the interview with the „Tageszeitung" (taz), 18/08/2003, here according to Cuttitta 2004, p.134

Melilla. He is holding a Master of Economic Science in European Economic and Public Affairs. He urges on the one hand that the roots ("root causes") of migration, poverty and conflicts have to be dealt with and on the other hand O'Connell warns that at the short term further incidents at the outer EU borders have to be avoided and the human rights have to be respected by the border control officials.

As a measure of a global reception of illegal migration he also sees the take back treaties and less hindrance for elite migration. A far going strategy "on the external dimension of the area of freedom, security and justice" of the EU commission from the 12th of October 2005[109] aims at bringing obedience to laws, democratic values and healthy authorities to the developing countries. The developing countries should get support from the EU also on the field of synergies between bank transfers back to the home countries by emigrants and the development cooperation sector. O'Connell draws attention to the fact, that the tackling of the economic root causes of migration probably needs the time of generations of people. Because of this it is asked from the EU to begin to work openly with connections of fields of policies especially the bad effects of the common agriculture policies to the African economies.[110] In my opinion the latter seems to be a very important aspect. As long as there is no justice concerning the world trade the developing countries cannot find a way out of their miserable financial situation. Since many years the developing countries have been asking already for the ending of hindrances for the export of agricultural products to the EU – without any solution.

[109] European Commission communication, "A Strategy on the External Dimension of the Area of Freedom, Security and Justice", 12th of October 2005, available in the internet under: http://www.europa.eu.int/comm/justice_home/news/intro/doc/com_2005_491_en.pdf, here according to O'Connell 2005, without given pages, from the internet as mentioned above

[110] O'Connell 2005, without given pages, from the internet, filed under: http://www.monitor.upeace.org/archive.cfm?id_article=315, 03/01/2008

What Next?

If the concept of a lawful state building and of human rights shall be introduced successfully worldwide, the respecting of human rights has to be enforced within the EU, because only this way the EU can keep being a good moralistic example for the whole world, as O'Connell makes clear in his conclusion.

Cuttitta does not allow any optimism regarding to the above mentioned while he is writing:

> "The integration of human rights affairs to the trans Mediterranean relationships will rather be problematic for international cooperation in the future according to the opinion of many politicians. They are afraid of restrictions through those human rights affairs for the enforcement of border control policies that they want to achieve."[111]

I see that as a confirmation for my hypothesis that they build a "Fortress of Europe" and produce problems for the lives of illegal migrants.

It keeps being worth hoping that the governments rethink their policies. Useful methods could be seen in the policy of the "export" of the building of lawful states and the improvement of the development cooperation sector. But as long as there is no consequent and real ending of the debt service of the developing countries and we keep holding up the false and unjust world trade policies it is difficult to believe in good propositions or alternating in the thinking.

Maybe the wish of the Reggae musician Tiken Jah Fakoly from Ivory Coast becomes fulfilled in the future and the youth in Europe awakes and raises its voice loudly against the political corruption. This would be in my own interest, anyway. Fakoly in the documentary "Au clair de la lune…":

> "You keep your borders closed while you yourself come to our countries as if you would use a toilet! You come whenever you want to, ask for a visa and the next day you get it. They do not ask you too many questions. I do not find this just, and I hope that the youth in the western countries awakes!"[112]

[111] Cuttitta 2004, p.134

[112] see in the appendix « 'Au clair de la lune...' - … » … Transcription of Subtitles, here translated into English from the German subtitle.

Sources

Monographic Books and Compendiums

Bade, Klaus J. (2000): Europa und die Migration am Ende des 20. Jahrhunderts. Akademievorlesung gehalten am 4. Juli 2000 (=Berichte aus den Sitzungen der Joachim Jungius-Gesellschaft der Wissenschaften e.V., Hamburg, Year 18, Vol. 5), Göttingen.

Currle, Edda (2004): Migration in Europa – Daten und Hintergründe (= europäisches forum für migrationsstudien (efms): Forum Migration, Vol. 8, Bamberg), Stuttgart.

Guild, Elspeth / Paul *Minderhoud* (Ed., 2006): Immigration and Criminal Law in the European Union. The Legal Measures and Social Consequences of Criminal Law in Member States on Trafficking and Smuggling in Human Beings (= *Guild*, Elspeth / Jan *Niessen* (Hg.): Immigration and Asylum Law and Policy in Europe, Vol. 9), Leiden, Boston.

Heckmann, Friedrich / Wolfgang *Bosswick* (Ed., 1995): Migration Policies: a comparative perspective (= europäisches forum für migrationsstudien (efms): Forum Migration, Vol. 1, Bamberg), Stuttgart.

ICS (Consorzio Italiano di Solidarietà, 2005): La protezione negata. Primo rapporto sul diritto di asilo in Italia

„Le Monde diplomatique“ / taz Verlags- und Vertriebs GmbH (Ed., 2006): Atlas der Globalisierung. Die neuen Daten und Fakten zur Lage der Welt, Berlin.

Milborn, Corinna (2006): Gestürmte Festung Europa. Einwanderung zwischen Stacheldraht und Ghetto. Das Schwarzbuch, Wien, Graz, Klagenfurt.

Articles in Compendiums

Puggioni, Raffaela (2006): Looking for some coherence: Migrants in-between criminalisation and protection in Italy, in: *Guild*, Elspeth / Paul *Minderhoud* (Hg.): Immigration and Criminal Law in the European Union. The Legal Measures and Social Consequences of Criminal Law in Member States on Trafficking and Smuggling in Human Beings (= *Guild*, Elspeth / Jan *Niessen* (Ed.): Immigration and Asylum Law and Policy in Europe, Vol. 9), Leiden, Boston, pp.169-200.

Ventrella McCreight, Matilde (2006): Crimes of assisting illegal immigration and trafficking in human beings in Italian law: Illegal immigration between administrative infringement and criminal offence, in: *Guild*, Elspeth / Paul *Minderhoud* (Hg.): Immigration and Criminal Law in the European Union. The Legal Measures and Social Consequences of Criminal Law in Member States on Trafficking and Smuggling in Human Beings (= *Guild*, Elspeth / Jan *Niessen* (Ed.): Immigration and Asylum Law and Policy in Europe, Vol. 9), Leiden, Boston, pp.141-168.

Zincone, Giovanna (1995): Immigration to Italy: Data and Policies, in: *Heckmann*, Friedrich / Wolfgang *Bosswick* (Ed.): Migration Policies: a comparative perspective (= europäisches forum für migrationsstudien (efms): Forum Migration, Vol. 1, Bamberg), Stuttgart, pp.137-156.

Articles in Journals

Cuttitta, Paolo (2004): „Africa ad portas? Italiens Migrationspolitik”, in: "Welt-Trends. Zeitschrift für internationale Politik und vergleichende Studien“ no. 45, pp.124-134

Kopp, Karl (2007): „Hart an der Grenze“, in: „Tag des Flüchtlings 2007“, pp.32-33, From the

Internet at the 21/12/2007, filed under:
http://www.proasyl.de/fileadmin/proasyl/fm_redakteure/Archiv/HTdF_Hefte/Webversion_Heft_komplett.pdf
Klepp, Silja (2007): „Libyen – Partner für den europäischen Flüchtlingsschutz?“, in: „Tag des Flüchtlings 2007“, p.37, from the Internet at the 21/12/2007, filed under:
http://www.proasyl.de/fileadmin/proasyl/fm_redakteure/Archiv/HTdF_Hefte/Webversion_Heft_komplett.pdf

Böffgen, Kerstin (2007): „Endstation Grenzzaun“, in: „Tag des Flüchtlings 2007“, pp.40-41, from the Internet at the 21/12/2007, filed under:
http://www.proasyl.de/fileadmin/proasyl/fm_redakteure/Archiv/HTdF_Hefte/Webversion_Heft_komplett.pdf

Articles in Email-Newsletters

(2003) „Italien: Hilfe bei Grenzsicherung gefordert“, exclusively filed in the Internet, in: „Migration und Bevölkerung“ no. 9, from the internet at the 19/10/2007, filed under:
http://www.migration-info.de/migration_und_bevoelkerung/artikel/030906.htm

(2003) „Mittelmeerraum: Wieder zahlreiche Bootsflüchtlinge gelandet“, nur im Internet, in: „Migration und Bevölkerung“ no. 6, from the Internet at the 15/10/2007, filed under: http://www.migration-info.de/migration_und_bevoelkerung/artikel/030608.htm

(2005) „Aufstände auf Lampedusa“, in: „Migration und Bevölkerung“ no. 6, „Kurzmeldungen - Europa“, p.4, from the Internet at the 01/01/2008, filed under:
http://www.migration-info.de/migration_und_bevoelkerung/artikel/050608.htm

(2005) „Marokko / Spanien: Massive Einreiseversuche in Exklaven“, in: „Migration und Bevölkerung“, no. 9, p.1-2, from the Internet at the 25/12/2007, filed under:
http://www.migration-info.de/migration_und_bevoelkerung/archiv/ausgaben/ausgabe0509.pdf

(2006) „EU: Kriseninterventionsteams und gemeinsame Patrouillen“, in: „Migration und Bevölkerung“ no. 6, p.3, from the Internet at the 06/01/2008, filed under:
http://www.migration-info.de/migration_und_bevoelkerung/archiv/ausgaben/ausgabe0606.pdf

(2007) „EU: Illegale Migration per Boot hält an“, in „Migration und Bevölkerung“ no. 1, pp.3-5, from the internet at the 05/01/2008, filed under:
http://www.migration-info.de/migration_und_bevoelkerung/archiv/ausgaben/ausgabe0701.pdf

(2007) „EU: Initiativen in den Bereichen Migration und Grenzschutz“, exclusively filed in the internet, in: „Migration und Bevölkerung“ no. 5, from the Internet at the 17/10/2007, filed under: http://www.migration-info.de/migration_und_bevoelkerung/artikel/070508.htm

Peers, Steve (no given year): „Readmission Agreements and EC External Migration Law“, in: “Statewatch analysis” no. 17, from the Internet at the 10/01/2008, filed under:
http://www.statewatch.org/news/2003/may/readmission.pdf

Pro Asyl (Ed., 2005): „ECRE bemängelt EU-Reaktion auf die Ereignisse in den spanischen Enklaven“, in: Newsletter 105, no given pages, from the Internet at the 21/12/2007, filed under:
http://www.proasyl.de/de/informationen/newsletter-nr-105/index.html#c821

Internet (Except Articles in Journals or Email-Newsletters)

Amnesty International, branch in the Federal Republic of Germany, statement for the press: „Zum Tag des Flüchtlings. Neues Gutachten: Flüchtlingsrechte gelten auch im Mittelmeer", from the Internet at the 20/12/2007, filed under: http://www.proasyl.de/fileadmin/proasyl/fm_redakteure/Europa/Frontex/Fluechtlingstag_27Sep07__2_.pdf

Burkhardt, Günter: „Auszug aus dem Statement zum Pressegespräch am 27. September 2007 in Berlin", from the Internet at the 20/12/2007, filed under: http://www.proasyl.de/fileadmin/proasyl/fm_redakteure/Europa/Frontex/Statement_Pressegespraech_am_27_September__2_.pdf

European Commission (2004): „Technical Mission to Libya on Illegal Immigration 27 Nov – 6 Dec 2004. Report", from the Internet at the 07/01/2008, filed under: http://www.statewatch.org/news/2005/may/eu-report-libya-ill-imm.pdf

European Commission communication (2005): "A Strategy on the External Dimension of the Area of Freedom, Security and Justice", 12 October 2005, under: http://www.europa.eu.int/comm/justice_home/news/intro/doc/com_2005_491_en.pdf

Fischer-Lescano, Andreas / Tillmann *Löhr* (2007): legal report „Rechtsgutachten. Menschen- und flüchtlingsrechtliche Anforderungen an Maßnahmen der Grenzkontrolle auf See", from the Internet at the 20/12/2007, filed under: http://www.proasyl.de/fileadmin/proasyl/fm_redakteure/Europa/Frontex/Gutachten_Geltung_MRe_auf_See_Sept._2007.pdf

Fortress Europe (Ed., 2007): "Escape from Tripoli. Report on the Conditions of Migrants in Transit in Libya", from the Internet at the 21/12/2007, filed under: http://www.statewatch.org/news/2007/nov/fortress-europe-libya-report.pdf

Frontex (2007): "Frontex-led EU Illegal Immigration Technical Mission to Libya 28 May-5 June 2007", from the Internet at the 21/12/2007, filed under: http://www.proasyl.de/fileadmin/proasyl/fm_redakteure/Newsletter_Anhaenge/130/eu-libya-frontex-report.pdf

Frontex / Finance, from the Internet at the 21/12/2007, filed under: http://www.frontex.europa.eu/finance/

Frontex / Legal basis, from the Internet at the 21/12/2007, filed under: http://www.frontex.europa.eu/legal_basis/

Gleitze, Judith / Alice *Schultz* (2006): „Zonen der Rechtlosigkeit. Eine Reise auf den Spuren der Flüchtlinge durch Süditalien", from the Internet at the 20/12/2007, filed under: http://www.proasyl.de/fileadmin/proasyl/fm_redakteure/Broschueren_pdf/Zonen_der_Rechtlosigkeit_04.pdf

Gleitze, Judith / Alice *Schultz*: „In den Zonen der Rechtlosigkeit. Italiens Asylsystem ist geprägt von Willkür, Rechtsverweigerung und Isolation – oder – Eine Reise durch Süditalien auf den Spuren afrikanischer Flüchtlinge", from the Internet at the 20/12/2007, filed under: http://www.proasyl.de/de/italien/italiens-asylsystem/index.html

Commission of the European Communities (2005): „Grünbuch über die Zukunft des europäischen Migrationsnetzes", from the Internet at the 08/12/2007, filed under: http://ec.europa.eu/justice_home/doc_centre/immigration/docs/com_2005_606_de.pdf

O'Connell, Kevin John (2005): "Fortress Europe: Ceuta and Melilla", from the Internet at the 03/01/2008, filed under: http://www.monitor.upeace.org/archive.cfm?id_article=315

Council of the European Union: Vorentwurf einer Antwort auf die schriftliche Anfrage E-

3720/07 von Angelika Beer, Betr.: „Sammelcharterflüge zur Durchführung von Abschiebungen“ from the Internet at the 06/01/2008, filed under: http://www.proasyl.de/fileadmin/proasyl/fm_redakteure/Newsletter_Anhaenge/127/PE-QFRONTEX.doc

Speech of the German minister of the interior Wolfgang Schäuble at the 19th of November 2007, from the Internet at the 21/12/2007, filed under: http://www.bmi.bund.de/nn_662956/Internet/Content/Nachrichten/Reden/2007/11/BM__Euro Med.html

Speech of Franco Frattini (conclusion) at the 29th of September 2006 in Madrid, from the Internet at the 08/12/2007, filed under: http://ec.europa.eu/commission_barroso/frattini/archive/speech_29_09_06_en.pdf

Texeire, Fiona (2006): „At The Gate of Fortress Europe: Irregular Immigration and Malta“, from the Internet at the 22/12/2007, filed under: http://www.alternattiva.org.mt/filebank/documents/Fiona%20TEXEIRE-%20At%20the%20Gate%20of%20Fortress%20Europe.pdf

UNHCR (2007): „Anmerkungen von UNHCR zu diplomatischen Zusicherungen und Internationalem Flüchtlingsschutz“, from the Internet at the 06/01/2008, filed under: http://www.proasyl.de/fileadmin/proasyl/fm_redakteure/Newsletter_Anhaenge/123/Diplomatische_Zusicherungen.pdf

www.united.non-profit.nl/pdfs/deathlist2005.pdf, without any given date

Email

„E-Mail der Europaabgeordneten Gabi Zimmer nach dem Besuch im Lager Lampedusa am 28.6.05“, in: Flüchtlingsrat Brandenburg (Ed.): Italia-Info 16/9/2005, without given pages, from the Internet at the 21/12/2007, filed under: http://www.proasyl.de/fileadmin/proasyl/fm_redakteure/Newsletter_Anhaenge/105/Italien_info.pdf

Other Sources

Conclusions of the Euro-Mediterranean Mid-Term Meeting of Foreign Ministers, Crete, 26-27 May, 2003

Conclusions of the Euro-Mediterranean Conference of Ministers of Foreign Affairs, Naples, 2-3 December, 2003

Consorzio Italiano di Solidarietà (2000): “Dossier Nausicaa. Primo Quadro sulla Tutela del Diritto d’Asilo in Italia”, unpublished paper

Relevé des conclusions de la Présidence and Proposition de mise en œuvre de la Déclaration de Tunis la 2ème Conférence Ministérielle sur la Migration en Méditerranée Occidentale (Rabat, les 22 et 23 Octobre 2003)

Map Footage

Pro Asyl (Ed., without any given year): “Europa. Mit den wichtigsten Grenzübergängen“

Pictures

Picture 1, page 8:

According to interviews with Alain Morice (Department „Immigration et société“, Centre national de la recherche scientifique (CNRS), Paris) as well as with Olivier Clochard (Migrinter, Piontiers) 2005

Picture 2, page 10:

Map from Olivier Clochard (Migrinter, Poitiers) and « Le Monde diplomatique » / Sources: United (www.unitedagainstracism.org) / Noborder (www.noborder.org/dead.php) / Migreurop (www.migreurop.org) / Association des familles de victimes de l'immigration clandestine (Afvic)

Picture 3, page 11:

Photo: Alice Schultz, source: *Gleitze*, Judith / Alice *Schultz* (2006): Zonen der Rechtlosigkeit. Eine Reise auf den Spuren der Flüchtlinge durch Süditalien, from the Internet at the 20/12/2007, filed under:

http://www.proasyl.de/fileadmin/proasyl/fm_redakteure/Broschueren_pdf/Zonen_der_Rechtlosigkeit_04.pdf

Picture 4, page 18:

Map from Olivier Clochard (Migrinter, Poitiers) / Data from numerous sources in the network Migreurop / UNHCR 2005 / Stefano Liberti, „il manifesto“

Pictures 1, 2 and 4 here according to Philippe Rekacewicz, from:

„Le Monde diplomatique“ / taz Verlags- und Vertriebs GmbH (Ed., 2006): Atlas der Globalisierung. Die neuen Daten und Fakten zur Lage der Welt, Berlin.

Appendix

Documentary «Au clair de la lune...», Transcription of the German Subtitles

„'Au clair de la lune...' – Realitès françafricaines", Documentary Movie, Burkina Faso, Ivory Coast, Mali, Morocco, 2005/06, with English, French and German Subtitles, Plot, Director and Camera: Leona Goldstein

Some Subtitles and Writings on Dark Background (translated by the author):

"Ivory Coast is the world's biggest producer of cocoa. Also coffee and petroleum made the country an 'area of interest'. In 2005 75 percent of the overall production profits went to French enterprises."

"In the past few years the Brussels based EU Commission spent about one billion dollars a day for agriculture subsidies. As a consequence in West Africa European vegetables average a third cheaper than local products. Estimations say that the prolonged EU agriculture subsidies will contribute to the disappearance of two third of the African agricultural area and that another 135 million people will be fleeing until the year 2025."

Halidou Ouedraogo, president of the U.I.D.H. (Inter-African human rights association) in the documentary:

"You have a country here where the judiciary system has got a problem. The 'Collective against impunity' in which 65 organizations of the civil society are united demand the clarification of almost 200 crimes. From somewhere this gets blocked.

We do not live democracy here in Burkina. We at most are in a process of democratization. Instead of 'democratic' I would call the country an institutionalized military state."

Tara Topan, president of Kebayina (Women's Movement in Burkina Faso):

"The one who talks about democracy hast to talk about freedom, too. If you look on Burkina Faso you could think that there is absolute freedom. But in my eyes it is a two-faced freedom. Because if you take the liberty of expressing yourself critical about the regime, as soon as you are becoming bothersome for the rulers, they will not hesitate to eliminatc you."

"Norbert Zongo. Main spokesman of the intellectual opposition movement and editor of the critical weekly journal 'L'Independant'. At the time of his death he investigated a murder in which Francois Compaore, the brother of the president, was involved. On 13th of December 1998 Norbert Zongo was stopped on his way home and burnt in his car."

„In 2005 the public development aid of the western industrialized countries to the developing countries added up to 58 billion dollars. In the same year the developing countries had to pay 482 billion dollars as debt service to the industrialized countries."

"In total there are 40 million people fleeing worldwide. About 32-34 million of those stay in their regions of origin. In Europe less than 200,000 people applied for asylum in 2006. On European average 5 percent of them became accepted."

Morocco/Spain (contribution from PRODEIN): "We hide. Often we move only at night. During the day we look for hiding places in the shadow of trees so that no one can see us. Because if they find us they often call for the authorities."

An African man with one foot wounded lies at the wayside, another one who is not fully shown by the camera arrives. – "Guardia… Si Guardia Civil"

The colored man reports: "I was about to jump across the border as they came with the weapons, they beat us and shot at us. We were 3. 4 military men and 3 Africans. The others fell in a faint. One oft hem died this night. He lies over there. He is dead. They beat as like animals. With the rears of their guns. They broke my foot. I cannot walk any more."

Voiceover: "For the 'security of the EU external borders against unwanted immigration' Morocco received about 40 million Euros from the EU between 2004 and 2006. Another 70 million flow from 2007 onwards."

Bandou Keita, 52 years old, another migrant who got repelled in Morocco at the Spanish border, reports: "These have been bad days, the authorities did not mean well with the refugees. They really wanted to kill us. Other comrades have already been abandoned in the Sahara. Just arrived at the police station they asked me for name, age, nationality. Meanwhile I saw comrades who sat such… and such… [he shows how by sitting down on the ground with spraddled, stretched legs]. In one row. Only a little gap in between us so that the military could pass. In the police station of Nador I saw comrades who sat like this 5 days long. Day and night. Such they sat the whole night in that row. Even if you were exhausted and tried to bend your legs they beat you with sticks. There were injured people beneath us with broken arms and legs, sick persons. Often there was no distance between us at all. The whole night we spent this way.

Often there were comrades who asked me: 'Why are you so sad? After all we become transported to Oujda today, because we are so many. They will bring lorries today and bring us to the border.' I said: 'You just do not have a clue what is happening in Morocco at the moment! All your comrades who climbed on these lorries here were abandoned in the desert!' There were injured people who got their arms or legs broken and could not walk any more. As soon as they had been thrown to the Sahara they had to stay there and wait for their death."

Aminata Traoré, Bamako, writer and former culture secretary of Mali:

"The alternative is clear: So that Africa changes and does not become an emigration continent, firstly Europe has to change. Europe has to stop stealing and lying. To its own people, but first of all to the Africans. Europe has to become aware that as long as it does not get controlled the social consequences of the neoliberal policies in Europe itself it has no right to demand from the African countries to compensate the consequences of exactly this policy which itself enforces upon Africa. This is which is so scandalous!"

Tiken Jah Fakoly, musician from Ivory Coast, in exile in Bamako, Mali:

"You keep your borders closed while you yourself come to our countries as if you would use a toilet! You come whenever you want to, ask for a visa and the next day you get it. They do not ask you too many questions. I do not find this just, and I hope that the youth in the western countries awakes!"

About the author

Julian Traublinger was born in 1981 in Bad Reichenhall, Germany. In his childhood he already has been interested in politics, particularly human rights issues. At secondary school he founded a students' group at Amnesty International. He successfully finished his studies of geography in 2008. Alongside his studies at the university he was politically active in an ecological party as well as with the network of globalization critics ATTAC. During a stay in Malta he got inspired to write about boat refugees. There he was able to visit a refugees' camp with a delegation of the Maltese Green Party.